D0835542

This delightful book is the latest in the series of Ladybird books which have been specially planned to help grown-ups with the world about them.

As in the other books in this series, the large clear script, the careful choice of words, the frequent repetition and the thoughtful matching of text with pictures all enable grown-ups to think they have taught themselves to cope. The subject of the book will greatly appeal to grown-ups.

Series 999

THE LADYBIRD
BOOKS FOR GROWN–UPS SERIES

THE
NERD

by

J.A. HAZELEY, N.S.F.W. and J.P. MORRIS, O.M.G.

(Authors of 'I Think You'll Find There's Only One Road
In Britain On Which You Are Legally Obliged To Drive On
The Right And I Think You'll Find It's Outside The Savoy
Hotel In London As A Matter Of Fact')

Publishers: Ladybird Books Ltd., Loughborough
Printed in England. If wet, Italy.

This is a nerd.

He has a PhD and works in the biochemistry lab of a university hospital.

But this is not what makes him a nerd.

His catalogued and indexed collection of over 6,000 empty crisp packets does that.

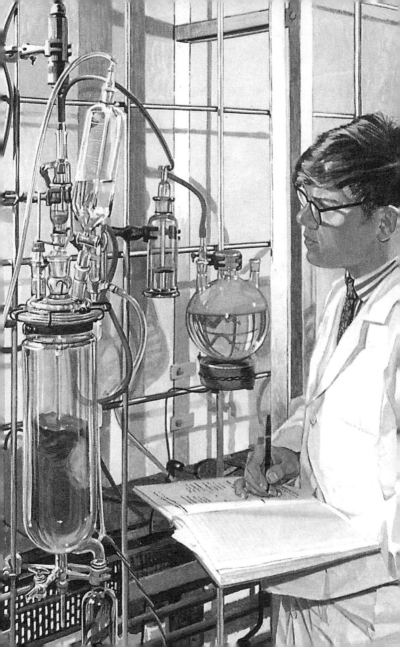

Wilbur has been waiting all his life to see this long-lost Doctor Who story "Inheritance Of The Worgons".

The episodes have been lovingly restored so that the sound and image are as good as they were in 1969.

Unfortunately, so are the plot, budget and pacing.

The BBC plans to lose all the recent series of Doctor Who so that nerds will like them too.

Gwaedlys is determined to win an argument about the name of Aethelflaed's second—favourite horse. She has spent four hours in the library trying to find a source that agrees with her, but has had no luck.

She will now start an argument with the librarians about the quality of their stock. She will not win that one either.

This comic meet is taking place in a convention centre in Leicester. There is only creamflow bitter on tap, or Foster's.

Ogden, Winston and Juliet are smuggling bottles of craft ale in to the centre under cover of darkness.

If the other convention goers spot them, they will be killed for their booty.

Everyone at work knows that Helen is a Star Wars nerd.

"My big thing Star Wars is, hmm? Yes? That was my Yoda impression! Excuse the nerd!"

Star Wars is one of the defining cultural phenomena of the last hundred years and has been seen by over a billion people.

Oliver has been painting this Elven Paladin miniature for the last six months.

He looks at the rest of the unpainted chess set. There are an awful lot of pawns.

This is a toy shop.

It has models of robots and superheroes. There are cars and dollies and fun things about popular television programmes and cartoons and films.

You can tell it is a toy shop and not a hobby shop because the members of staff do not have any piercings and their t−shirts do not look as though they have been through the wash 1,000 times.

Jenna has given up writing cyberpunk fan fiction because the field is too crowded and the fan comments are too mainstream.

Her 15th Century wickerpunk reboot of Firefly is proving to have a reassuringly niche appeal.

"The latest OS update has a few bugs," says Nico. "And I don't much like the new font," says Tom. "Me neither," says Toby.

Craig says nothing. He still uses Netscape Navigator 1.0 because it has "never been bettered," he will insist, if you ask him, which you shouldn't.

His weekly blog about still using Netscape Navigator 1.0 is available by mail order, as Netscape Navigator 1.0 will no longer connect to the internet.

For his art project, Garf has taken the jpegs of his weekend in Snowdonia and transferred them to colour slides. He will then project the slides on to a screen and film the results on a vintage 16mm cine camera, before choosing still frames from the developed film, scanning them and showing the finished jpegs on an iPad.

Garf tends to eat yellow food.

Amanda is making costumes for World Book Day at Viola and Harry's school.

"If anyone asks," Amanda tells the children, "say you are Wasp from the Avengers and Violator from the Spawn universe."

Amanda has a well-rehearsed speech about how comics count as books.

Someone will probably enjoy this.

emulsion sensitive to
blue light only

yellow filter to prevent blue light
reaching lower layers

emulsion sensitive to green light

gelatine

emulsion sensitive to red light

film base

SECTION OF COLOUR FILM

THE ORIGINAL COLOUR SCALE

Section of emulsion. After exposure to the colourscale and first
development a negative silver image is formed
in each layer.

The remaining silver salt is exposed and colour developed
producing both dye and silver positive images in each layer.

All the silver images are removed leaving only the positive
dye images.

On projection the three images together reproduce
the original colour scale.

THE REVERSAL PROCESS (Colour Transparencies)

Horatio has been up all night correcting the Wikipedia page for a 1968 Gerry Anderson TV pilot that was wiped in 1973.

He is happy to see that the page is now slightly longer than the page for the Crimean War.

Wanda and Ed met at work. They share many enthusiasms, like the music of They Might Be Giants and the film Napoleon Dynamite.

One night they sit by the river watching shooting stars. Wanda remarks on how romantic it is.

Ed replies by talking for fifteen minutes about meteor showers, especially the Quadrantids and the Leonids. He is very clever.

Ed will never notice the way Wanda looks at him.

Gibby is at his brother—in—law's house fixing the broadband again.

Gibby has tried to explain that just because he is really into J.R.R. Tolkien, that does not mean he knows anything about computers, but his brother—in—law tends to glaze over after the second "R".

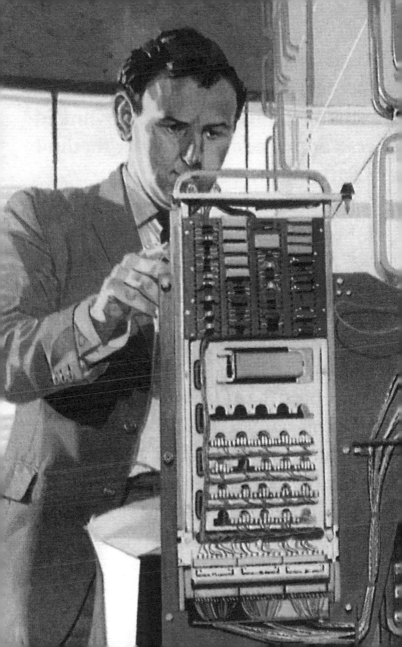

The Leyton Buzzard Modelling Collective are building a working model of a BattleMech from an anime series that was recently cancelled.

It is not for a convention or a competition. They are planning to take over the Earth.

They will then force the channel to make more episodes.

From the age of four, Zoë insisted to her teachers that she did not want to learn cursive script, as she preferred her handwriting to be in Helvetica.

Des and Ludwig met in 2007 on a Big Bang Theory fan forum.

Since then, they have not said more than eleven words to each other that have not been quotes from the programme.

Des and Ludwig are the best of friends.

Travis is at a science—fiction convention. He is dressed as the audiobook version of Paul McGann. Everyone at the convention recognises his outfit and smiles.

He shows a photo of his costume to some people at work. "Who are you meant to be? Him off Changing Rooms?" they ask.

Travis wishes that being at conventions was his job.

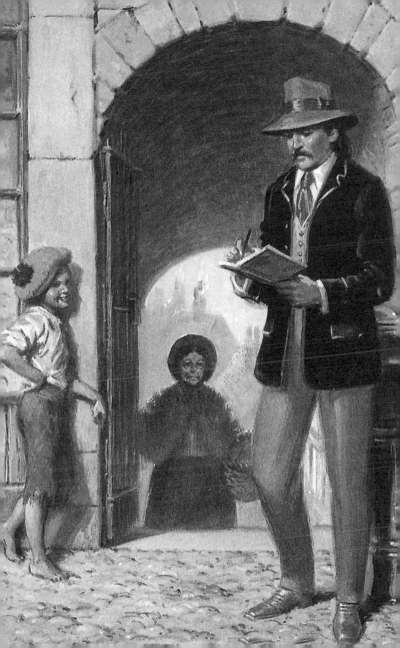

Denholm is going on the Million Apostrophe March.

The March was meant to take place three months ago, but it has taken some time for all the placards to be proof-read by a council of grammar elders.

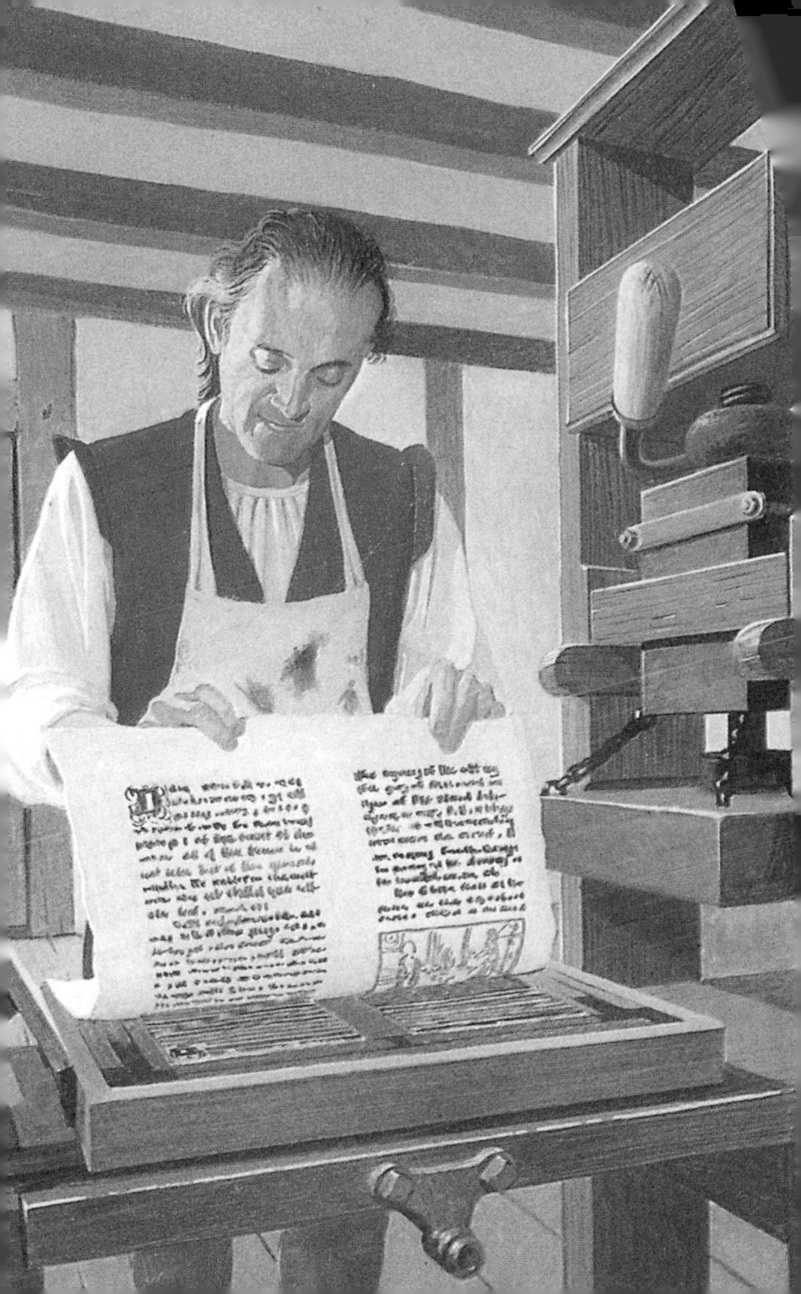

Roddy is a nerd. He likes Star Trek and programming his Raspberry Pi, but does not like football.

He will be mercilessly teased for the next ten years by people who will eventually earn a quarter of what Roddy earns at CERN.

Revenge is a dish best served cold. Roddy knows this. It is an old Klingon proverb.

These nerds have paid money to see a film they think is terrible and have seen at least thirty times before.

They are having the most fun they can possibly have.

Nerds don't just like science fiction. They like science too.

This electron microscope has been designed to detect the tiny, imperceptible differences between the plots of Star Wars: A New Hope and Star Wars: The Force Awakens.

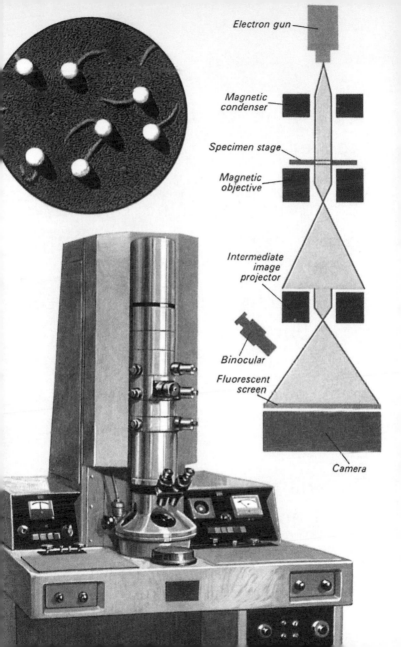

Electron gun

Magnetic condenser

Specimen stage

Magnetic objective

Intermediate image projector

Binocular

Fluorescent screen

Camera

"Count up to ten," says Ant's teacher.

"Eight, five, four, nine, one, seven, six, ten, three, two," says Ant.

The teacher tries to correct him, but Ant points out that the numbers are in alphabetical order, just like the letters in his name are.

She will probably not ask Ant to count up to a hundred.

THE AUTHORS would like to record their gratitude and offer their apologies to the many Ladybird artists whose luminous work formed the glorious wallpaper of countless childhoods. Revisiting it for this book as grown-ups has been a privilege.

MICHAEL JOSEPH

UK | USA | Canada | Ireland | Australia
India | New Zealand | South Africa

Michael Joseph is part of the Penguin Random House group of companies whose addresses can be found at global.penguinrandomhouse.com

First published 2017
001

Copyright © Jason Hazeley and Joel Morris, 2017
All images copyright © Ladybird Books Ltd, 2017

The moral right of the authors has been asserted

Printed in Italy by L.E.G.O. S.p.A

A CIP catalogue record for this book is available from the British Library

ISBN: 978-0-718-18864-1

www.greenpenguin.co.uk

MIX
Paper from
responsible sources
FSC® C018179

Penguin Random House is committed to a sustainable future for our business, our readers and our planet. This book is made from Forest Stewardship Council® certified paper.